SCRIPTURE WORKBOOK

Our Catholic Identity®

Grade Four

Scripture Advisor
Rev. Ronald J. Nuzzi, Ph.D.

Authors
Jim Bitney
Mary Beth Jambor
Yvette Nelson
Dee Ready

RCL
Resources for Christian Living®
Allen, Texas

President: Kim Duty
Publisher: Maryann Nead
Catechetical Advisor: Jacquie Jambor
Editorial Director: Ed DeStefano

Scripture Advisor: Rev. Ronald J. Nuzzi, Ph.D.
Authors: Jim Bitney, Mary Beth Jambor, Yvette Nelson, Dee Ready

Contributor: Judy Deckers

Product Manager: Kate Sweeney Ristow
Senior Project Editor: Anne Battes Kirby
Senior Editor: Patricia A. Classick
Editor: Linda Hartley
Senior Production Editor: Laura Fremder

Art Director: Pat Bracken
Art Coordinator: Karen Malzeke-McDonald
Photoshop Artist: Carol-Anne Wilson
Page Design: Laura Fremder
Production Manager: Jenna Nelson
Cover Design: Carol-Anne Wilson, Karen Malzeke-McDonald
Cover Art: Margaret Sanfilippo

NIHIL OBSTAT
Rev. Msgr. Glenn D. Gardner, J.C.D.
Censor Librorum

IMPRIMATUR
† Most Rev. Charles V. Grahmann
Bishop of Dallas

January 12, 1998

The Nihil Obstat and Imprimatur are official declarations that the material reviewed is free of doctrinal or moral error. No implication is contained therein that those granting the Nihil Obstat and Imprimatur agree with the contents, opinions, or statements expressed.

ACKNOWLEDGMENTS

Scripture excerpts are taken from the *New American Bible with Revised New Testament and Psalms* Copyright © 1991, 1986, 1970 Confraternity of Christian Doctrine, Washington, DC. Used with permission. All rights reserved. No part of the *New American Bible* may be reproduced by any means without the permission of the copyright owner.

Photos: The British Museum, 64B; Sonia Halliday Photographs/Jane Taylor, 60B; The Israel Museum, 60T; © 1998 Loyola University Chicago, R. V. Schoder, S.J., photographer, 64T; Bill Wittman, 35.

Illustrations: Pamela Johnson, 8, 9, 10, 11, 12, 13, 14, 15, 20, 21, 30, 31, 34, 35, 54, 55, 58, 59; Paula Lawson, 48, 49; Debra McClatchey-Ames, 26, 27; Bob Niles, 6, 7, 16, 17, 52, 53; Margaret Sanfilippo, 18, 19, 22, 23, 24, 25, 28, 29, 32, 33, 34, 35, 36, 37, 38, 39, 40, 41, 42, 43, 44, 45, 46, 47, 49; Schechner and Associates, 25; Elizabeth Swisher, 50, 51, 56, 57.

Send all inquiries to:
RCL • Resources for Christian Living
200 East Bethany Drive
Allen, Texas 75002-3804

Toll Free 800-822-6701
Fax 800-688-8356

Visit us at www.rclweb.com

Printed in the United States of America

20264 ISBN 0-7829-0851-9 (Student Book)
20274 ISBN 0-7829-0892-6 (Annotated)

2 3 4 5 6 07 06 05 04 03

Contents

Part Four: The Works of Jesus

Part Five: The Last Days of Jesus on Earth

Part Six: The Spirit of Jesus and The Church

Scripture Glossary

WELCOME

Welcome to *Our Catholic Identity*® Scripture Series. This book has stories from the Bible to help you learn about the Word of God. In each Scripture story there is a special message for the people of God—us!

Follow these steps when reading each chapter in the book:

1. Begin by learning about the people in the Scripture story in the BIBLE BACKGROUND.
2. Read the story in READING THE WORD OF GOD.
3. Learn what the story means, or its message, through UNDERSTANDING THE WORD OF GOD.
4. Find out some interesting facts about the people, places, and terms IN BIBLE TIMES.
5. Complete each ACTIVITY to help remember the Scripture story and its message.
6. Share your thoughts about the message in the story in GOD'S WORD IN MY LIFE.
7. Memorize a SCRIPTURE VERSE to help you remember God's Word.
8. Apply or learn more about a Scripture story WITH FAMILY OR FRIENDS.

Scripture Glossary, the section at the end of the book, provides vocabulary words of people, places, and terms. They are divided into the two main parts of the Bible, the New and Old Testaments. Each part has a map to show you where the Scripture stories took place.

Journey with the people in the Old Testament to discover God's love for all people. Continue the journey with Jesus and the followers of Jesus to learn how God's love changes the lives of people. Enjoy your journey through Scripture and learn how God's message helps us today!

1
The Bible
MATTHEW 5:17–20

Bible Background

The **Bible,** or **Sacred Scripture,** is the written Word of God. The Bible has two main parts. The first part of the Bible is called the **Old Testament.**

The Old Testament tells about God's people who lived before **Jesus** was born. The Old Testament contains forty-six books. These books are divided into the **Law,** or Torah; the **Prophets;** and the **Writings.**

Reading the Word of God

Read **Matthew 5:17–20.**

Discover what Jesus says about the Old Testament.

Understanding the Word of God

In this story, many people respect Jesus as a teacher. Jesus helps people understand the Old Testament. People know that God's law in the Old Testament shows them how to live as God's people. They are to live as children of God. The prophets help the people follow God's ways. Jesus' words and actions teach about God's love.

The next four stories are from the Old Testament. Read about some of God's chosen people.

IN BIBLE TIMES

The books of the Old Testament were handwritten by scribes. The word *scribe* means "writer."

Learning about the Bible

Solve the code to find an important message about the Old Testament.

Morse Code

A	·–	G	––·	M	––	S	···	Y	–·––
B	–···	H	····	N	–·	T	–	Z	––··
C	–·–·	I	··	O	–––	U	··–		
D	–··	J	·–––	P	·––·	V	···–		
E	·	K	–·–	Q	––·–	W	·––		
F	··–·	L	·–··	R	·–·	X	–··–		

__ __ __ __ __ , __ __ __ __ __ __ __ __ __
·––– · ··· ··– ··· – · ·– –·–· ···· ·· –· ––· ···

__ __ __ __ __ __ __ __ __ __
· –··– ·––· ·–·· ·– ·· –· – ···· ·

__ __ __ __ __ __ __ __ __ __ __ __
–– · ·– –· ·· –· ––· ––– ··–· – ···· ·

__ __ __ __ __ __ __ __ __ __ __ .
––– ·–·· –·· – · ··· – ·– –– · –· –

God's Word In My Life

As a child of God, I . . .

"I have come not to abolish but to fulfill."
Matthew 5:17

With Family or Friends

Make a special place for the Bible in your home. Open the Bible to your favorite story and share it.

2
Jacob
GENESIS 27:1–29

Bible Background

Abraham was the first great leader of God's people. God made a promise, or a **covenant,** with Abraham. God promised Abraham that he would be the "father of many nations."

Abraham and **Sarah** had only one son, **Isaac.** Isaac and his wife, **Rebekah,** had twin sons, **Esau** and **Jacob.**

Reading the Word of God

Read **Genesis 27:1–29.**

Find out what happens to Jacob.

Understanding the Word of God

In this story, the son who receives the blessing of the father becomes the head of the family. Jacob receives his father's blessing and becomes the new head of the family. God fulfills his covenant with Abraham through Jacob.

In Bible Times

Jacob had twelve sons. They became the leaders of twelve groups of God's people. The Bible calls these groups the twelve tribes of Israel.

Remembering the Story of Jacob

Write the name of each person in the list under his or her description. Then complete the family tree.

Abraham **Jacob** **Esau** **Isaac**

I am the first leader of Israel.

____

I am the father of Jacob and Esau.

____

Jacob was my younger twin brother.

____

I became the new head of Isaac's family.

____

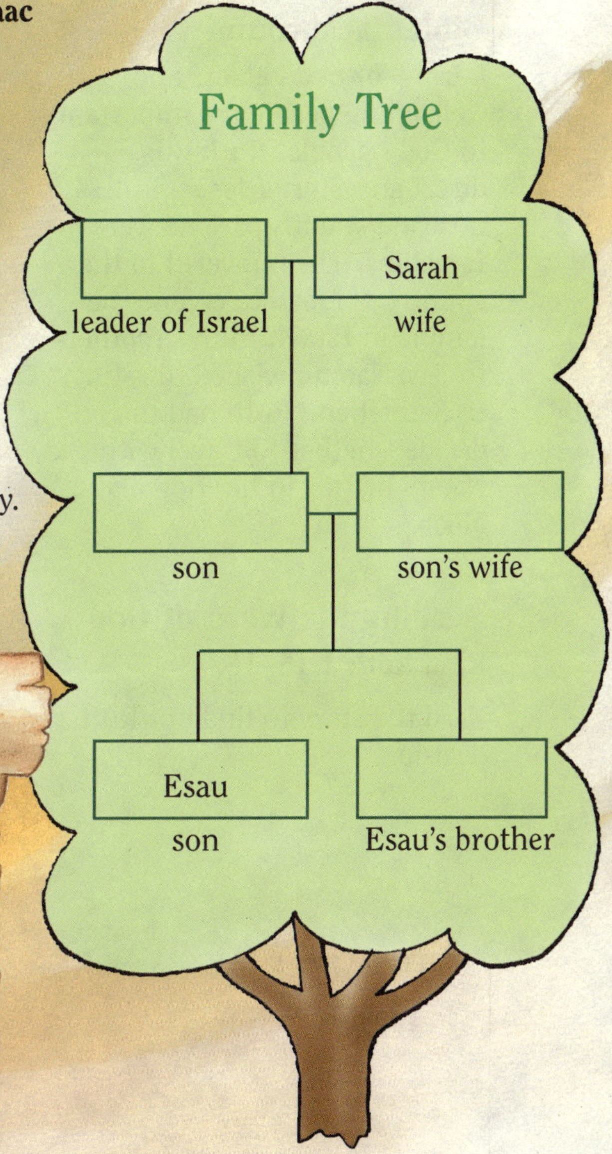

God's Word In My Life

God is with me . . .

"Let peoples serve you, and nations pay you homage." GENESIS 27:29

With Family or Friends

Make a "Hall of Heroes" poster. Use photographs of friends and family members who help you live as a child of God.

3
Ruth and Naomi
Ruth 1:15–18

Bible Background
Caring for one's elders and welcoming people are important to God's people. **Ruth** was a direct ancestor of Jesus. Ruth's husband and his parents were **Israelites.** They traveled to the country of **Moab** to escape the famine in **Israel.** Ruth's mother-in-law, **Naomi,** wished to return to Bethlehem. Ruth had to decide whether to travel with Naomi or stay in her homeland, Moab.

Understanding the Word of God
The story of Ruth teaches that God is merciful and loving to all people. Ruth is a kind and hospitable woman. She is generous to people. She chooses to take care of her mother-in-law, Naomi. Ruth leaves her family and the land of her birth. She lives with Naomi among the people in Israel. The hospitality of Ruth is a sign of God's love for all people.

Reading the Word of God
Read **Ruth 1:15–18.**

Read the story to find out Ruth's decision.

In Bible Times

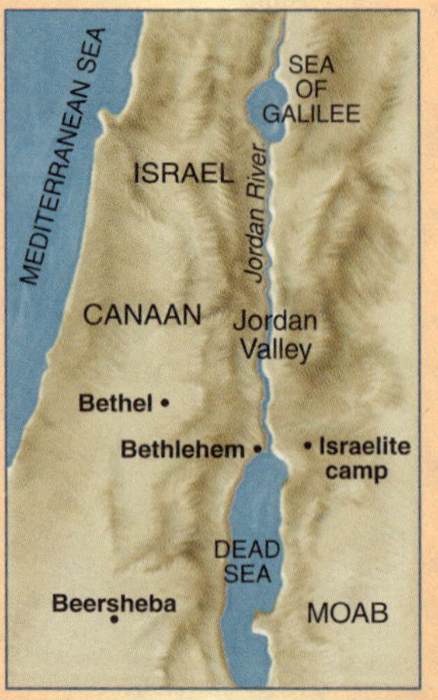

Recalling Ruth's Words

Find the words of Ruth in the Bible. Draw a line to connect the sentence parts that go together. Read the sentences aloud and think about what God is saying to you.

1. Wherever you go,	**a.** I will lodge.
2. Wherever you lodge,	**b.** is my God.
3. Your people	**c.** I will go.
4. Your God	**d.** shall be my people.

God's Word In My Life

I love all God's people when . . .

"Your people shall be my people."
Ruth 1:16

With Family or Friends

Write a prayer using Ruth's words. Ask God to help you accept and respect everyone. Treat them with kindness. Pray your special prayer.

4
The Call of Samuel

1 SAMUEL 3:1–10, 19–20

Bible Background

Samuel was one of the **"judges"** of Israel. There were fifteen judges named in the Bible. The judges were the leaders of God's people before they had kings. They were filled with God's spirit.

Samuel's mother was **Hannah.** She prayed to God for a child. God answered Hannah's prayer. When Samuel was little, his mother brought him to the shrine at Shiloh. The priest, Eli, taught Samuel as he grew up.

Reading the Word of God

Read **1 Samuel 3:1–10** and **19–20.**

Find out how Samuel learns about the Lord.

Understanding the Word of God

In this story, Samuel hears a voice calling him. He thinks it is Eli, the priest. Eli helps Samuel understand that the voice he hears is the Lord.

The story shows Samuel's willingness to listen to God. Samuel says, "Speak, for your servant is listening" (1 Samuel 3:10).

IN BIBLE TIMES

The names of people in the Bible tell about the relationship between God and the person. Hannah named her son Samuel. The name means "one who hears God."

Discovering Samuel's Message

Solve the rebus. The second word uses the sound that the picture makes. Each word in the riddle answer has its own rebus. Below the fourth word, write Samuel's response to God's call.

Word 1: טלביטוחדגבא – brew +

 – b, ad = ____________

Word 2:  = ____________

Word 3:  – ltar +  – ary = ____________ ,

Word 4:  – amb +  – ination = ____________ .

"__ __ __ __   __   __ __ ,   __ __ __ __ __ ."

With Family or Friends

Make a list of ways you can help one another listen to God. Practice the ideas on your list often.

God's Word In My Life

God, I am listening to you when . . .

"Speak, Lord, for your servant is listening."
1 Samuel 3:9

5
Saul and David
1 Samuel 26:6–25

Bible Background

The Israelites trusted and believed that God was their only true king. However, they were surrounded by powerful nations that were led by kings. So the Israelites wanted a "human" king, too. **Saul** became the first king of God's people.

Reading the Word of God

Read **1 Samuel 26:6–25.**

Find out what happens to Saul and **David,** another future king of God's people.

Understanding the Word of God

In this story, King Saul becomes jealous of a good soldier named David. Saul takes his army to find David and kill him. **Jonathan,** the son of King Saul, tells David about King Saul's plan.

Even though David had a chance to kill Saul, he chooses not to kill him. Instead, David honors the king as a person specially chosen by God. Saul later tells David that he was wrong and will never harm David. King Saul and David show the people how to honor God.

In Bible Times

Priests used oil to anoint kings. The best oil came from not-yet-ripened olives. The priest poured the olive oil over the king's head as a sign of being specially chosen by God.

Illustrating the Story of Saul and David

Create a Table of Contents page for a book using the story of Saul and David. Write the title and the names of the first three chapters.

With Family or Friends

Remember that you were anointed with oil at baptism. Tell how you are a leader because of your baptism.

____

Title

Table of Contents

Chapter 1

____

Chapter 2

____

Chapter 3

____

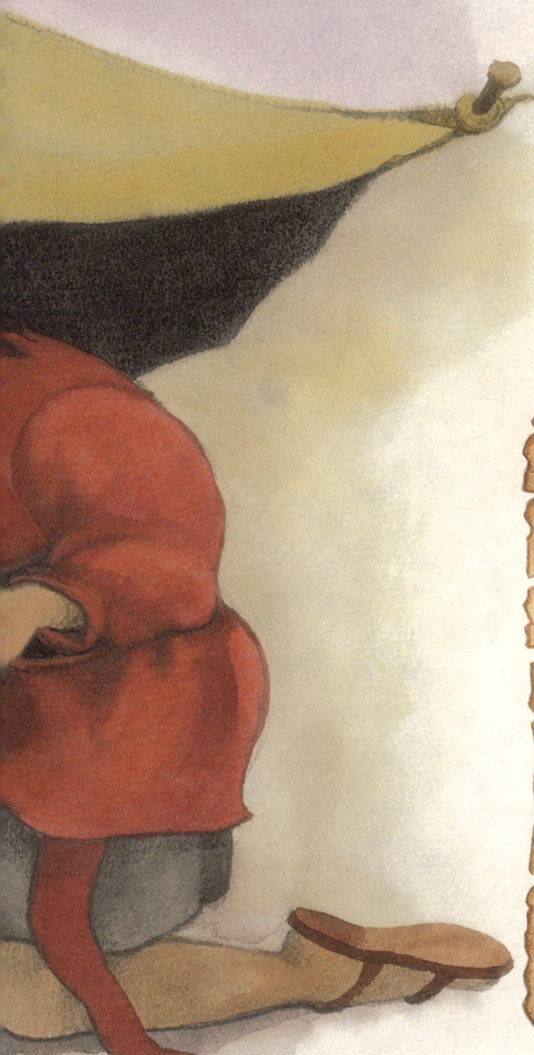

God's Word In My Life

God, I honor you when . . .

"Blessed are you."
1 Samuel 26:25

6
The New Testament

ACTS OF THE APOSTLES 1:1–5

Bible Background

The **New Testament** is the second main part of the Bible. It is the written Word of God. It contains twenty-seven writings. Four of these writings are **Gospels.** *Gospel* means "good news."

The Gospel writers are Matthew, Mark, Luke, and John. They pass on what the first Christians believed about the life, death, resurrection, and ascension of Jesus.

The New Testament also contains the **Acts of the Apostles** and twenty-one letters to Christian people. The last book of the Bible, **Revelation,** tells about Jesus coming again.

Reading the Word of God

Read the **Acts of the Apostles 1:1–5.**

Read what Luke writes about the Holy Spirit.

Understanding the Word of God

Luke begins the Acts of the Apostles with the promise of the Holy Spirit. Luke tells how the Holy Spirit will help the Christian community grow. The Holy Spirit helps people learn and understand about Jesus.

The rest of the stories in this book are from the New Testament. Read what Jesus said and did about the early Church.

IN BIBLE TIMES

Luke, the evangelist, was a writer and may have been a physician. The word *evangelist* means "a writer of the good news." Luke came from the city of Antioch and traveled with Paul the Apostle.

Praying to the Holy Spirit

Learn this prayer to the Holy Spirit. Pray it often.

Come, Holy Spirit, fill the hearts of your faithful.
And kindle in them the fire of your love.
Send forth your Spirit, and they shall be created.
And you will renew the face of the earth.

Let us pray.

Lord,
by the light of the Holy Spirit
you have taught the hearts of your faithful.
In the same Spirit
help us to relish what is right
and always rejoice in your consolation.

We ask this through Christ your Lord.
Amen.

God's Word In My Life

Come, Holy Spirit, help me . . .

"You will be baptized with the holy Spirit."
ACTS OF THE APOSTLES 1:5

With Family or Friends

Write the prayer to the Holy Spirit on a card. Pray it with your family.

7
The Annunciation

LUKE 1:26–38

Bible Background

God sent an angel to a young Jewish girl, **Mary.** The word *angel* means "messenger of God." The angel **Gabriel** greeted Mary, saying, "The Lord is with you" (Luke 1:28).

Reading the Word of God

Read **Luke 1:26–38.**

Read the story to find the good news that Gabriel tells Mary.

Understanding the Word of God

In this Gospel story, God sends the angel Gabriel to Mary. Gabriel tells Mary that God has chosen her to be the mother of God's Son. Mary says yes to God's love for her.

Mary's yes is an act of faith and love. Mary believes and trusts that God will help her be the mother of God's Son.

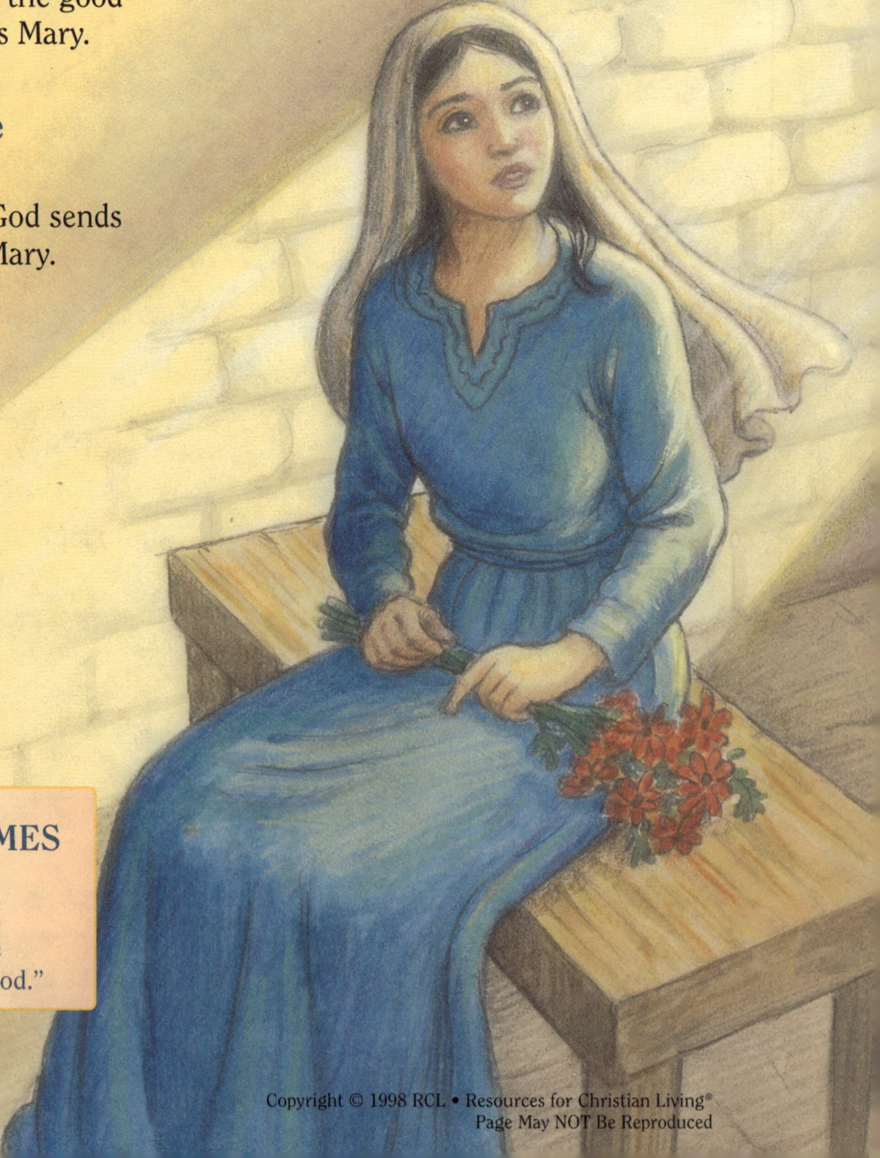

IN BIBLE TIMES

The Bible mentions several angels by name. The angel Gabriel visited Mary. The name *Gabriel* means "champion of God."

Learning about Mary

Complete the sentences about Mary. Then write the words into the puzzle.

yes	faith	believed	God	Mary	Gospel	mother

1. ______________ sent an angel to visit Mary.
2. ______________ was chosen to be the mother of God's Son.
3. Mary ______________ that God would help her.
4. Mary will become the ______________ of God's Son.
5. When Mary said yes, it was an act of ______________ .
6. Mary said ______________ to God.
7. You can read the story of Mary in Luke's ______________ .

G __ __

__ A __ __

B __ __ __ __ __ __ __

__ __ __ __ __ R

__ __ I __ __

__ E __

__ __ __ __ __ L

God's Word In My Life

Yes, God, I will . . .

"Behold, I am the handmaid of the Lord."
Luke 1:38

With Family or Friends

Pray the Hail Mary. The first part of the Hail Mary tells about Mary being chosen to be the mother of Jesus, God's Son.

8
Songs of Praise
LUKE 1:39–80

Bible Background
Mary traveled to visit her cousin **Elizabeth.** She wanted to help Elizabeth when her baby was born. She sang a **canticle** of praise to God. Canticle is another word for "song." In the Bible, a canticle is a song of wonderful praise to God.

Reading the Word of God
Read **Luke 1:39–80.**

Read how these canticles "sing" praises to God for the wonders he has done.

Understanding the Word of God
Mary's canticle is called the **Magnificat.** In this canticle, Mary gives praise for all God has done for her. The canticle reminds us of God's loving mercy.

The canticle of **Zechariah** is called the **Benedictus.** This canticle "blesses" God for caring for all people.

IN BIBLE TIMES
There are four canticles in the Gospel according to Luke. They are Mary's canticle, Zechariah's canticle, the angels' canticle, and Simeon's canticle.

Singing God's Praises

Luke's Gospel contains both Mary's and Zechariah's song of praise to God. Look up these two verses in your Bible. Find out which verse is Mary's and which one is Zechariah's. Write the correct name on the line.

1. ____________________________ sang,
"My soul proclaims the greatness of the Lord; my spirit rejoices in God my savior." **Luke 1:46–47**

2. ____________________________ sang,
"Blessed be the Lord, the God of Israel, for he has visited and brought redemption to his people." **Luke 1:68**

Write an opening verse for your own canticle of praise to God. Write your verse here:

____

____

____

God's Word In My Life

Praising God is . . .

"My spirit rejoices in God my savior."
Luke 1:47

With Family or Friends

List ways you praise God throughout the week. We can praise God by what we say and what we do.

9
The Birth of Jesus

LUKE 2:1–7

Bible Background

Bethlehem is called the "City of David." Israel's greatest king, David, was born in Bethlehem. He was born about one thousand years before the birth of Jesus. God promised King David that a **Messiah** would come from his family. The word *messiah* means "anointed one."

The birth of Jesus is the story of the Incarnation. The word *incarnation* means "to take on flesh."

Christians believe that the Son of God "took on flesh and became human." Jesus is both fully God and fully human.

Reading the Word of God

Read **Luke 2:1–7.**

Find out how Jesus, born in Bethlehem, fulfills God's promise.

Understanding the Word of God

The birth of Jesus reveals that God loves us and keeps his promise to us.

IN BIBLE TIMES

Tradition says that Joseph walked and Mary rode a donkey from Nazareth to Bethlehem. The distance from Nazareth to Bethlehem is about ninety miles.

Discovering Jesus' Birth

Explain how Jesus, born in Bethlehem, fulfills God's promise. Use the words in the box to help you. Draw or write your answer.

David	**census**	**Jesus**
Messiah	**Bethlehem**	**God**

With Family or Friends

Act on a promise you made to a family member or friend.

God's Word In My Life

Jesus' birth helps celebrate . . .

She gave birth to her firstborn son.

Luke 2:7

10
The Presentation of Jesus

LUKE 2:22–38

Bible Background

It was the religious custom of the Jews to present their firstborn son to God. Mary and **Joseph** dedicated, or presented, Jesus to God in the Temple.

Reading the Word of God

Read **Luke 2:22–38.**

Find out what happens when Jesus is presented in the Temple.

Understanding the Word of God

The presentation of Jesus shows the **Holy Family** obeying Jewish religious law. Mary and Joseph bring Jesus to the Temple. They offer two turtledoves instead of lambs.

Simeon and **Anna** meet the Holy Family.

Simeon has been waiting all his life to see the Messiah. Simeon takes Jesus in his arms, knowing he is holding God's Promised One. He tells Mary that Jesus will save God's people.

Anna has also been waiting. When she sees the Holy Family, she praises God for fulfilling his promises. She tells everyone in the Temple that the Messiah has come.

IN BIBLE TIMES

People sacrificed animals to give thanks and praise to God. Lambs were usually the animal sacrificed. Poor people sacrificed turtledoves.

Recognizing the Messiah

Reread the words and actions of Simeon and Anna. In each turtledove, write something that Simeon or Anna said or did that shows Jesus is the Messiah, God's Promised One.

God's Word In My Life

I am light when . . .

"My eyes have seen your salvation."
Luke 2:30

With Family or Friends

Simeon said that Jesus was "a light" for others. Do or say something to brighten another person's life.

11
Parable of Judging Others

LUKE 6:37–42

Bible Background

Jesus often used **parables** to help people understand his message. A parable is a story that teaches a lesson. The word *parable* means "comparisons." Jesus told parables to help his listeners compare his message to their lives.

Reading the Word of God

Read **Luke 6:37–42.**

Discover Jesus' message in this parable.

Understanding the Word of God

In this Gospel, Jesus warns his listeners not to judge others. They need to forgive and be generous. God will repay them for their kindness.

Then Jesus shares a parable with the people. Many of Jesus' listeners know about working with wood. The parable uses the image of a beam of wood and a splinter. It tells how we see the splinter, or fault, in someone else. We do not notice the beam, or our own faults. People need to treat others as God treats everyone.

IN BIBLE TIMES

The words "a good measure, pressed down, shaken together, running over" (Luke 6:38) were all used to describe a plentiful harvest.

Understanding Jesus' Message

Jesus compared splinters and a beam of wood to people's faults. Read the following story. How would you use splinters and a beam of wood to explain the message of judging others?

Draw or write your own parable that explains how to treat others fairly.

With Family or Friends

Share a story about a person who was judged unfairly. Discuss ways to avoid judging others.

God's Word In My Life

I am fair . . .

"Forgive and you will be forgiven."
Luke 6:37

12
Martha and Mary

LUKE 10:38–42

Bible Background

Jesus was traveling with his disciples. A woman named **Martha** invited Jesus and his disciples to her home. Martha and her sister **Mary** cared for Jesus in different ways.

Reading the Word of God

Read **Luke 10:38–42.**

As you read the story, notice what Jesus says about the two sisters.

Understanding the Word of God

This Gospel shows Martha busily preparing everything for Jesus and the other guests. Her sister Mary does not help. Instead, Mary listens to what Jesus is saying.

Martha becomes upset, thinking Mary should help her. She wants Jesus to tell Mary to work. Jesus' answer is not the one Martha expects. He tells Martha not to concern herself with all the work. The most important work is listening to Jesus' message.

IN BIBLE TIMES

Houses were lit by putting olive oil in a clay dish with a wick. When the wick was lit, the fire slowly burned the wick and olive oil.

Learning about Martha and Mary

Find the words about Mary and Martha in the puzzle and circle them. Then use some of the words to write the main idea of the story.

Mary	**Martha**	**Jesus**	**listen**	**message**
work	**women**	**welcome**	**sister**	**task**

T	B	L	C	R	T	L
D	W	M	Q	O	A	I
Z	J	E	S	U	S	S
U	W	O	I	M	K	T
M	E	S	S	A	G	E
A	L	Z	T	R	H	N
R	C	F	E	Y	K	D
T	O	E	R	J	D	P
H	M	W	O	R	K	G
A	E	W	O	M	E	N

____

____

God's Word In My Life

The Word of God . . .

"Listen, O distant peoples."
Isaiah 49:1

With Family or Friends

List several ways you listen to Jesus. Choose one item on your list and continue doing it.

13
The Great Feast
LUKE 14:15–24

Bible Background

Jesus told the parable of the Great Feast to religious leaders called **Pharisees.** The word *Pharisees* means "separated ones." These leaders lived strictly by the Jewish law. This separated them from other groups of people who did not live strictly by the law.

to explain God's invitation. The host in this parable represents God. The Pharisees represent God's Chosen People. The dinner, or feast, represents the kingdom of God. All people are invited to God's kingdom.

Reading the Word of God

Read **Luke 14:15–24.**

Read to find out what Jesus tries to tell the Pharisees.

Understanding the Word of God

In this Gospel, Jesus wants the Pharisees to understand that God loves everyone. Jesus uses the invitation to a dinner

IN BIBLE TIMES

It was customary to send a person to call the guests who had been invited to a dinner.

Inviting All People

Make a guest list of all the people that the host invited to the feast. Then write the invitation. In the Come Celebrate section, include Jesus' message.

Guest List

____
____
____
____
____
____
____
____
____

To: ______________________

From: ______________________

Come Celebrate: ______________________

____
____
____
____
____

With Family or Friends

Eat lunch at school with someone you have not eaten with before. Enjoy your guests as Jesus did when he ate meals with people.

God's Word In My Life

I welcome others when . . .

"Come, everything is now ready."
Luke 14:17

14
Parables about Being Lost
LUKE 15:1–32

Bible Background
Jesus told three parables about being lost. These parables show the joy of being found.

Reading the Word of God
Read **Luke 15:1–32.**

As you read, notice the rejoicing that takes place.

Understanding the Word of God
In this parable, God is compared to a shepherd, a woman who lost a coin, and a forgiving father. Jesus compares a sinner to a lost sheep. He explains that the shepherd rejoices when he finds his lost sheep.

Jesus also compares a sinner to a lost coin. The woman rejoices when she finds her coin.

Jesus tells a story about a prodigal son. The word *prodigal* means "wasteful with money." The prodigal son returns home and apologizes to his father. The father welcomes his son and celebrates.

Jesus teaches that God "rejoices" every time we ask for forgiveness and change our ways.

IN BIBLE TIMES
A shepherd's job was to lead the flock to fresh pasture and water. Shepherds also protected their sheep from wild animals and from getting lost.

Finding What Is Lost

Solve the maze to help the shepherd, the woman, and the father find what they lost.

GOD'S WORD IN MY LIFE

God, forgive me when . . .

"There will be rejoicing among the angels of God over one sinner who repents." LUKE 15:10

With Family or Friends

Tell about a time when you felt better after apologizing for something you did. Tell why you felt better.

15
The Rich Young Man

LUKE 18:18–25

Bible Background

A rich young man asked Jesus about eternal life. The man wanted to know what he must do to get to heaven.

Reading the Word of God

Read **Luke 18:18–25.**

Find out what Jesus tells the young man.

Understanding the Word of God

The man refers to Jesus as "Good Teacher." Jesus is often called **Rabbi,** a title that means "teacher." Jesus tells the rich young man that all goodness comes from God. Jesus' goodness comes from God.

Jesus compares wealth to a camel going through the eye of a needle. The camel symbolizes the "large" barrier that wealth can be to the rich man. Sometimes wealth can be a barrier to live as God's children.

IN BIBLE TIMES

Some writers believe that the eye of a needle refers to a narrow gateway into a town. It could have been an opening for merchants.

Living God's Message

Read about Mother Teresa. Write about someone you know who gives possessions or time to help people.

Mother Teresa

Mother Teresa formed a religious community in Calcutta, India, called the Missionaries of Charity. The community cares for the world's poorest people. Mother Teresa worked especially with the sick and dying. Mother Teresa's caring community looks after all the people of God.

____

helps people by ____________

____

____

____

____

____

____

____

With Family or Friends

Name one possession that would be difficult for you to give away. Share it with someone.

God's Word In My Life

God, I care about . . .

"It is easier for a camel to pass through the eye of a needle than for a rich person to enter the kingdom of God." Luke 18:25

16
The Rejection at Nazareth

LUKE 4:16–30

Bible Background

Jesus began to teach the good news. He returned to **Nazareth,** his hometown. In the **synagogue,** Jesus read aloud the scroll containing the writing of the prophet Isaiah.

Reading the Word of God

Read **Luke 4:16–30.**

Find out what Jesus announces in the synagogue.

Understanding the Word of God

In this story, the Jews gather in the synagogue on the **Sabbath** each week. They read the Scriptures and listen to the rabbis teach. They pray for the coming of the Messiah. Jesus is respected as a teacher, or rabbi. So the people invite him to read the Scriptures and teach about them. Jesus tells them that God has heard their prayers. Jesus says, "Today this scripture passage is fulfilled in your hearing" (Luke 4:21).

At first the people are "amazed" and "speak well" of Jesus. But soon they become "filled with fury" (Luke 4:28). They drive Jesus out of town.

The story fulfills the prophecies that the Messiah will be rejected by his own people. Jesus will be rejected many more times.

IN BIBLE TIMES

The word *synagogue* means "assembly." The synagogue was a meeting place for prayer and for Scripture study.

Discovering a Prophecy Fulfilled

Write or draw two messages Jesus is telling the people.

Message 1

Message 2

GOD'S WORD IN MY LIFE

Jesus, you are . . .

"Today this scripture passage is fulfilled in your hearing." LUKE 4:21

With Family or Friends

Share a reading from the Bible together at least once a week. Tell how the reading helps you live as a Christian.

17
Miracles of Healing
Luke 8:40–56

Bible Background
A **miracle** is a sign of God's power and presence with us. Jesus performed miracles to help people believe in God's loving presence.

Reading the Word of God
Read **Luke 8:40–56.**

Read about two miracles that happen to people of faith.

Understanding the Word of God
In the first story, **Jairus** is a man of faith. He has faith that Jesus can help his dying daughter. Jesus sees Jairus' faith and sets off to see his sick child.

At Jairus' house, the people say that the little girl is dead. Nothing can be done. Jesus works a miracle and gives the child back to her family.

In the second story, the sick woman is also a person of faith. People believe her sickness is a punishment for her sins. But the woman has faith in Jesus and touches his robe. Jesus tells her that her faith has made her well.

Both miracles show the presence of God's love at work among his people. Miracles invite people to trust and believe in God's love for them.

In Bible Times
The beds people slept on were usually simple cloth or woven mats. In the daytime, beds would serve as places for people to sit and talk to one another or to God.

Showing God's Love

Write a news report for one of the miracles. In the report include a Headline, Who, What Happens, and Why.

____
Headline

Who: ______________________________

What Happens: ______________________________

____

Why: ______________________________

____

____

With Family or Friends

Keep a notebook. Each day list ways you share God's love.

God's Word In My Life

My faith helps . . .

"Your faith has saved you; go in peace."
Luke 8:48

18
Jesus Feeds Five Thousand

LUKE 9:10–17

Bible Background

Many people began to follow Jesus. Luke writes that there were about five thousand people who have followed Jesus. Jesus welcomed and cared for them.

Reading the Word of God

Read **Luke 9:10–17.**

Find out how Jesus cares for people.

Understanding the Word of God

This story shows how Jesus reveals that God cares for everyone. The people have no food. Jesus takes five loaves and two fish in his hands. He looks up to heaven and calls upon God. Then Jesus blesses the food and gives it to the **disciples.** They share it with the people. All eat and are filled.

Jesus' actions are similar to his actions in the story of the **Last Supper.** The words *blessed, broke,* and *ate* appear in this story and the Last Supper. At the Last Supper, Jesus blesses the bread and gives it to the disciples as his body.

IN BIBLE TIMES

Unleavened bread is a flat bread, like pita bread. It was made from barley. Barley was a common crop because it could grow in poor soil.

Feeding the Crowds

Jesus turned two fish and five loaves of bread into enough food for a crowd. Complete the puzzle with words from the gospel story. The first letter "E" has been done for you.

```
  B
  L
J E S U S
  S
  S
  E
  S
```

With Family or Friends

Find a favorite recipe for making bread. Bake bread. Share it at a special meal.

GOD'S WORD IN MY LIFE

God bless . . .

"God's love endures forever."
PSALM 136:25

19
The Last Supper
LUKE 22:14–20

Bible Background

At **Passover** the Jewish people celebrate their passage from slavery in Egypt to freedom. Each year the Jewish people remember and celebrate this important event.

Jesus journeyed to **Jerusalem** to celebrate the Passover with his disciples. Jesus sent two of his **apostles** to look for a room where they could celebrate the Passover meal.

Reading the Word of God

Read **Luke 22:14–20.**

Discover what Jesus says and does at this special meal.

Understanding the Word of God

This is the story that tells about Jesus giving us the Eucharist. Before his death, Jesus celebrates Passover with his disciples. At the meal Jesus takes a piece of unleavened bread and breaks it. Then Jesus says, "This is my body, which will be given for you" (Luke 22:19).

After the meal, Jesus passes the cup of wine for them to drink. He says, "This cup is the new covenant in my blood, which will be shed for you" (Luke 22:20).

During this meal Jesus gives his body and blood to the disciples. Jesus asks them to share this meal with one another. He says, "Do this in memory of me" (Luke 22:19). The Church is still following Jesus' command today at the celebration of the Eucharist.

In Bible Times

The Passover meal began with a meal of roasted lamb, unleavened bread, wine, and bitter herbs.

Finding Jesus' Message

At the Last Supper, Jesus gave us the Eucharist. Solve the code to find out why Jesus has given us this sacrament.

Look at the example. The first symbol is [symbol]. Find this shape [symbol] on the grid. Notice that the dot is in the first position. This means that "G" is the letter. Work through the "O" and "D." Now, solve the code.

Example: G o d

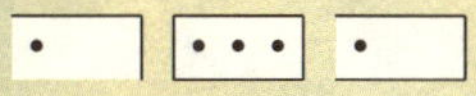

ABC	JKL	STU
DEF	MNO	VWX
GHI	PQR	YZ

Jesus said,

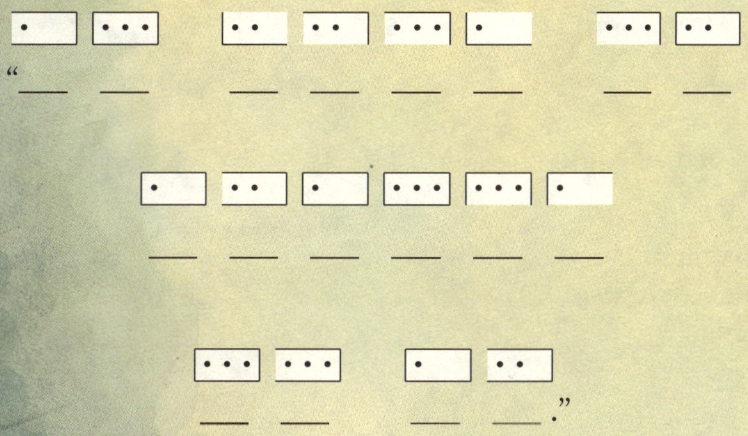

God's Word In My Life

Thank you, God . . .

"This is my body, which will be given for you."
Luke 22:19

With Family or Friends

Become an expert on a Passover meal. Read and discuss what takes place at the Passover meal.

20
The Garden of Gethsemani
LUKE 22:39–46

Bible Background
After the Passover meal, Jesus and his apostles walked to a garden at the foot of the **Mount of Olives.** A few days before, Jesus entered the holy city of Jerusalem. People were lining the streets and rejoicing. His procession into Jerusalem on that day began on the Mount of Olives.

Reading the Word of God
Read **Luke 22:39–46.**

Discover how this night is different from the day Jesus entered Jerusalem.

Understanding the Word of God
The Gospel of Luke teaches that Jesus prays often. He prays at all the important moments in his life. Jesus goes to the Mount of Olives into the **Garden of Gethsemani** "as was his custom" (Luke 22:39). That is, he goes there often to pray when he is near Jerusalem.

Jesus knows that the end of his life is near. In prayer, he asks God to "take this cup" from him (Luke 22:42). The "cup" means the difficult task ahead of him. But Jesus knows God will be with him. He accepts God's will.

In Bible Times
The Mount of Olives lies east of the city of Jerusalem. The Kidron Valley separates the Mount of Olives from the mount on which Jerusalem sits. The Garden of Gethsemani lies east of the Kidron Valley, at the base of the Mount of Olives.

Recalling the Actions of Jesus

Complete each of the five statements below. Select and circle the correct answer from the list underneath each statement.

1. After the Last Supper, Jesus and his apostles went to ______ .

   Jerusalem   the Mount of Olives   Bethlehem

2. At the base of the Mount of Olives is the Garden of ______ .

   Gethsemani   Kidron Valley   Jerusalem

3. The Mount of Olives lies to the east of Jerusalem across the ______ .

   Jordan River   Sea of Galilee   Kidron Valley

4. In the garden, Jesus ______ .

   slept   talked to the owner of the garden   prayed

5. Jesus asked his Father ______ .

   to send him more help   to teach the apostles to stay awake
   for the strength to do the Father's will

With Family or Friends

Read Psalm 27:1, 34:4–6, and 69:1–3. Choose the one you like best. Print it on paper. Pray it when you need help.

God's Word In My Life

God, I pray . . .

"Still, not my will but yours be done."
LUKE 22:42

21
Peter Denies Jesus
Luke 22:54–62

Bible Background

At the Last Supper, Jesus told **Peter** that he would deny Jesus three times. Jesus said that Peter would do this before the rooster crowed. In the dark of night, a crowd came and arrested Jesus.

Reading the Word of God

Read **Luke 22:54–62.**

Find out how Jesus' prediction about Peter comes true.

Understanding the Word of God

In this story, after the arrest of Jesus most of the disciples run into the night. Peter remains loyal even though he is afraid. He follows Jesus at a distance.

Peter's accent gives him away as being a Galilean, like Jesus. Bystanders accuse Peter of being Jesus' friend. Peter denies that he even knows Jesus. The third time Peter denies knowing Jesus, he hears the rooster crow.

Just then Jesus comes out. He turns and looks at Peter. Suddenly Peter remembers Jesus' prediction. Peter's denial helps people understand how much God forgives.

In Bible Times

If the night grew cold, people made fires outside. They hollowed a hole, put stones around it, and burned straw, grass, or charcoal.

Learning More about Peter

Using this word map, write four facts you know about Peter. Then write a paragraph describing his character.

Peter

God's Word In My Life

God, help me forgive . . .

"Lord, you know everything; you know that I love you."
John 21:17

With Family or Friends

Pray the Act of Contrition or write your own act of contrition. Pray it.

22
The Crucifixion and Burial of Jesus

LUKE 23:26–27, 33–38, 44–56

Bible Background

The religious leaders of Jerusalem took Jesus to **Pilate,** the Roman governor. The people in the crowd cried out, “Crucify, crucify him!” Pilate sentenced Jesus to death.

Reading the Word of God

Read **Luke 23:26–27, 33–38,** and **44–56.**

Follow Jesus as he carries his cross from Jerusalem to **Golgotha.**

Understanding the Word of God

Jesus carries his cross to the place of the skull called Golgotha. There Jesus is crucified by nailing him to the crossbeam. This is the cruelest form of punishment. Only the worst criminals are punished this way.

At the **Crucifixion** Jesus prays, “Father, forgive them, they know not what they do” (Luke 23:34). This is Jesus’ last great act of forgiveness.

At three o’clock Jesus gives up his spirit and dies. **Joseph of Arimathea,** a religious leader from Judea, offers to bury Jesus. He places Jesus’ body in a tomb cut out of a rock.

In Bible Times

Tombs were often cut in the soft limestone of a hill. A trench led to the entryway. Inside the tomb, the body lay on a niche in the stone wall. To close the tomb, mourners rolled a large stone in a groove they cut in front of the entryway.

Walking with Jesus

Christians today pray The Way of the Cross. The Way of the Cross helps us walk with Jesus on his journey from Jerusalem to Golgotha. In the box, draw another picture of Jesus' journey. Next to each picture, write a one-line prayer. Share your thoughts and feelings with Jesus.

God's Word In My Life

On my journey with Jesus . . .

"Father, forgive them, they know not what they do."
Luke 23:34

With Family or Friends

Learn more about the Way, or Stations, of the Cross. Visit your church and pray The Way of the Cross.

23
The Resurrection of Jesus

LUKE 24:1–35

Bible Background

After Jesus died, Joseph of Arimathea buried his body in a tomb. On the first day of the week, some women went to the tomb. When they went in, they did not find Jesus' body.

Reading the Word of God

Read **Luke 24:1–35.**

Try to imagine what the people in the two stories are feeling and thinking.

Understanding the Word of God

In the first story, two angels announce to the women that Jesus has risen. They remind the women that Jesus predicted his death and **resurrection.** The women are the first to learn of the Resurrection. They rush to tell the apostles, who do not believe them.

In the second story, the stranger is the risen Jesus. The two followers of Jesus are walking toward **Emmaus.** They do not recognize Jesus at first. They come to believe the stranger is Jesus when they "break bread" together. We believe and tell others the good news that Jesus is alive!

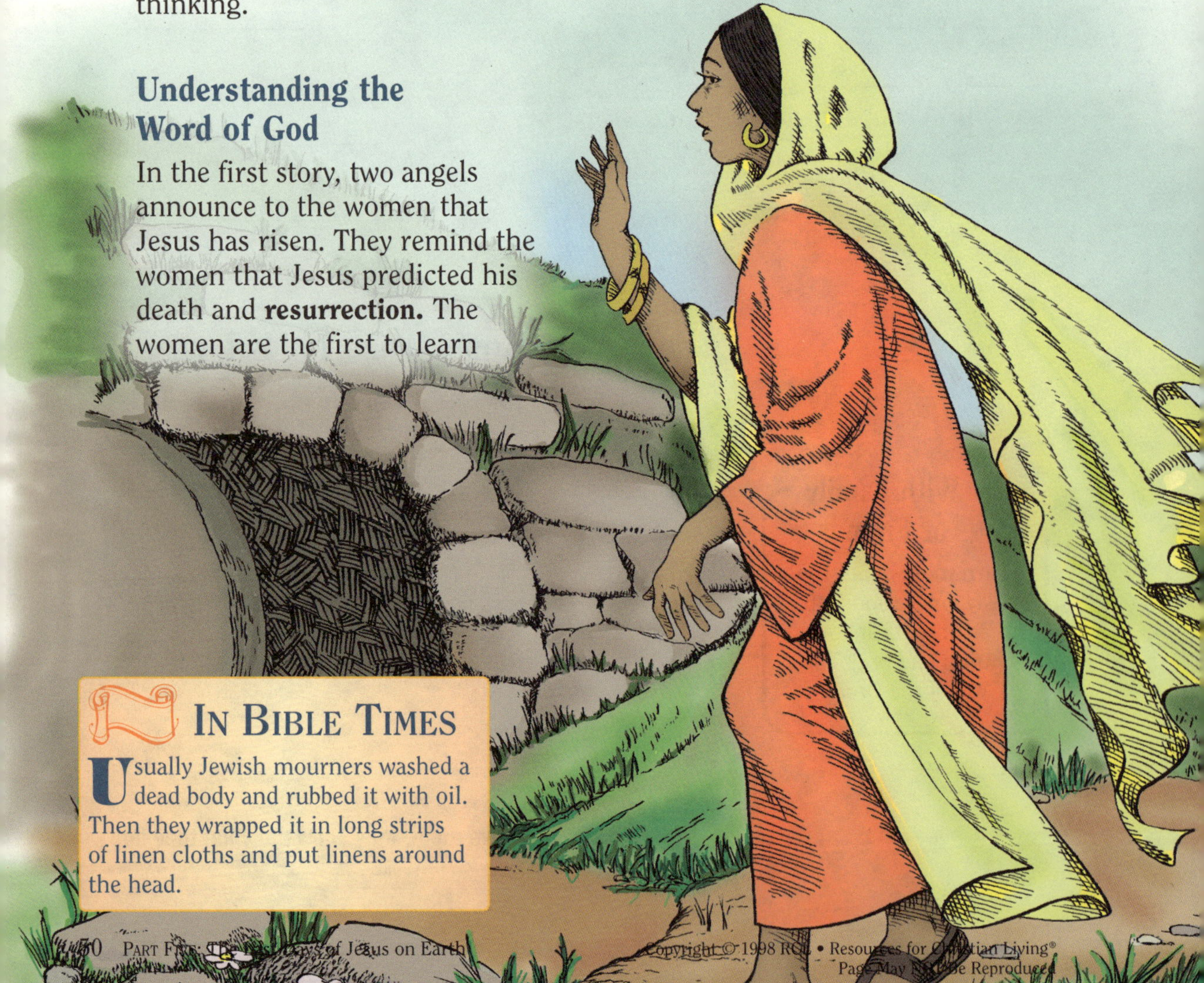

IN BIBLE TIMES

Usually Jewish mourners washed a dead body and rubbed it with oil. Then they wrapped it in long strips of linen cloths and put linens around the head.

Announcing the Resurrection

We celebrate our faith in Jesus' Resurrection on Easter Sunday. Write the ending for each part of the Easter story. Then write a headline for that part of the story.

1.

____

When the women arrived at the tomb early in the morning, they were surprised to find the stone rolled away from the tomb. Suddenly two men in dazzling clothes stood beside them.

____

____.

2.

____

As the two disciples were walking on the road to Emmaus, a "stranger" started walking and talking with them. The disciples invited the stranger to stay with them. As they ate their evening meal, the stranger took bread, blessed and broke it, and gave it to the disciples.

____

____.

God's Word in My Life

I recognize Jesus when . . .

"He is not here, but he has been raised."
Luke 24:6

With Family or Friends

Discuss the comparisons of the journey to Emmaus with our celebration of the Eucharist.

24
The Ascension of Jesus

LUKE 24:50–53

Bible Background

After his resurrection, Jesus appeared to many of his followers. He told them that he would send them the Holy Spirit. Jesus led them to Bethany outside Jerusalem.

Reading the Word of God

Read **Luke 24:50–53.**

Discover what happens in Bethany.

Understanding the Word of God

Christians believe that Jesus is the Messiah. He is the "high priest" of the kingdom of God. Like a high priest, Jesus blesses his disciples. He then ascends to heaven to his Father. This is called the **Ascension** of Jesus.

Then Mary and the disciples return to Jerusalem. Filled with joy, the followers of Jesus go to the Temple. There they thank God for his blessings. The greatest blessing of God is Jesus, the promised Messiah who lives forever.

In Bible Times

High priests announced God's will to the people. Some writers believe there were eighty-three high priests during the time of the Old and New Testaments. They wore special clothes. One piece of clothing had twelve jewels on it. The jewels represented the twelve tribes of Israel.

Discovering Ascension Words

Jesus ascended into heaven. Solve the crossword puzzle to discover words that remind you of Jesus' ascension.

ACROSS

2. The Temple is in the city of ______.
4. Jesus led his disciples to ______ outside Jerusalem.
5. Jesus ascended into ______.
6. ______ blessed his followers.

DOWN

1. As Jesus was ______ his followers, he was carried up into heaven.
2. The followers of Jesus were filled with ______.
3. The disciples blessed God in the ______.

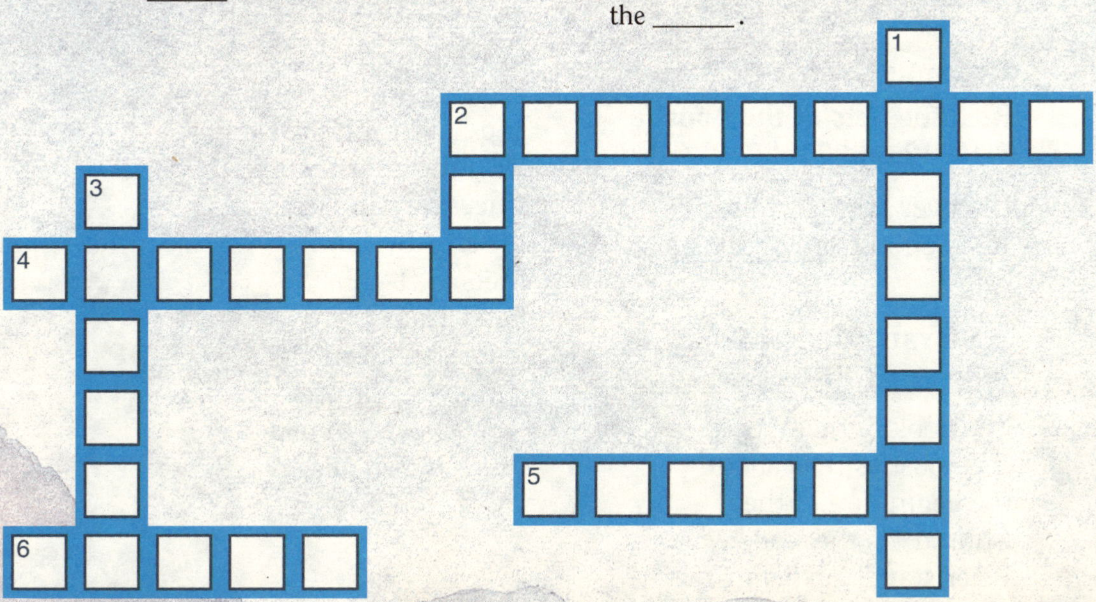

GOD'S WORD IN MY LIFE

Rejoice, O God . . .

"O LORD, my God, forever will I give you thanks."
PSALM 30:13

With Family or Friends

Interview your friends. Ask them what it means to be a follower of Jesus.

25
Matthias

ACTS OF THE APOSTLES 1:13–17, 21–26

Bible Background

After **Judas** died, there were only eleven apostles. They wanted to have twelve apostles again. It reminded them of the twelve tribes of Israel.

Reading the Word of God

Read the **Acts of the Apostles 1:13–17 and 21–26.**

Discover how the followers of Jesus choose a new apostle.

Understanding the Word of God

The book of the Acts of the Apostles tells about the beginning of the Church. As the leader, Peter stands up before the whole group. Then he suggests that someone needs to be chosen to take Judas' place as an apostle. Both Joseph and **Matthias** would be worthy apostles.

Luke tells what is required to be an apostle. The person chosen has to be a witness to Jesus' resurrection and have been with the apostles since the baptism of Jesus by John (Acts 1:21–22). Peter leads the whole group in prayer. Matthias is chosen to take Judas' place.

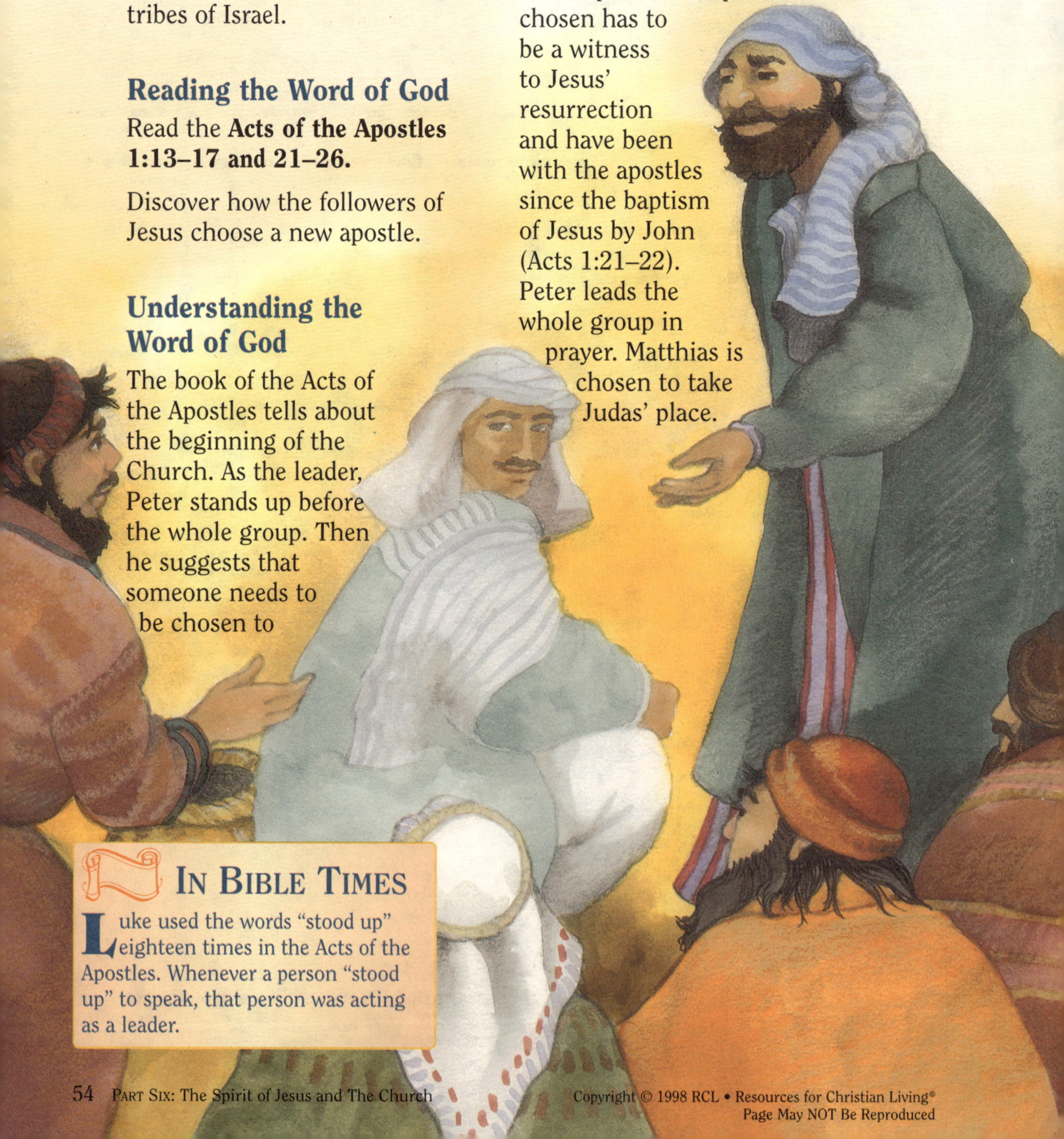

IN BIBLE TIMES

Luke used the words "stood up" eighteen times in the Acts of the Apostles. Whenever a person "stood up" to speak, that person was acting as a leader.

Recalling the Story

Read each statement. Draw a circle around the heads-up side of the coin for yes. Draw a circle around the tails-up side of the coin for no.

1. Peter was the leader of the apostles.

2. Peter wanted to forget about their Jewish ancestors.

3. Joseph and Matthias were both good choices to become an apostle.

4. One had to be a witness to Jesus' resurrection to be chosen to be an apostle.

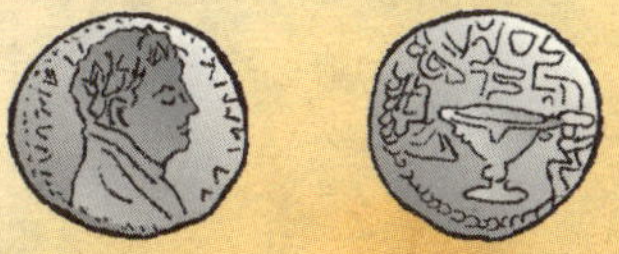

5. Matthias took Judas' place as an apostle.

God's Word In My Life

As your worthy follower . . .

"O God, hear my prayer."
Psalm 54:4

With Family or Friends

Become an expert on one of the apostles. Share your knowledge.

26
The Coming of the Holy Spirit

ACTS OF THE APOSTLES 2:1–41

Bible Background

Every year Jews from many countries gathered in Jerusalem to celebrate the feast of **Pentecost.** The Holy Spirit came upon the followers of Jesus on Pentecost.

Reading the Word of God

Read the **Acts of the Apostles 2:1–41.**

Discover how the coming of the Holy Spirit affects Peter and the disciples.

Understanding the Word of God

This story begins on the Jewish feast of Pentecost. The crowds are from many countries around the Mediterranean Sea. They speak many different languages. Filled with the Holy Spirit, Peter preaches to the crowds about Jesus. Preaching about Jesus is the main work of the apostles.

The apostles' preaching begins in Jerusalem and continues on to Rome. Jerusalem is the center of all Jewish life and worship. Rome is the center of the world at this time. The Acts of the Apostles teaches that Jesus came for the whole world. He came first to Jews and through them to all people.

IN BIBLE TIMES

Pentecost was a harvest festival. It occurred seven weeks, or fifty days, after the harvest of the first crops of wheat. On Pentecost, people offered a portion of their first harvest and two lambs to God as a sign of thanksgiving.

Discovering the Meaning of Pentecost

The first Pentecost marked the birthday of the Church. Solve the code telling what happened on that day.

A	B	C	D	E	F	G	H	I	J	K	L	M
1	2	3	4	5	6	7	8	9	10	11	12	13
N	**O**	**P**	**Q**	**R**	**S**	**T**	**U**	**V**	**W**	**X**	**Y**	**Z**
14	15	16	17	18	19	20	21	22	23	24	25	26

On the day of Pentecost, the ___ ___ ___ ___ ___ ___ ___ ___ were
1 16 15 19 20 12 5 19

filled with the ___ ___ ___ ___ ___ ___ ___ ___ ___ ___ .
8 15 12 25 19 16 9 18 9 20

___ ___ ___ ___ ___ ___ people from many countries heard the
10 5 23 9 19 8

___ ___ ___ ___ of the ___ ___ ___ ___ in their own native language.
23 15 18 4 12 15 18 4

Thousands were ___ ___ ___ ___ ___ ___ ___ ___ . They became
2 1 16 20 9 26 5 4

followers of ___ ___ ___ ___ ___ that day.
10 5 19 21 19

With Family or Friends

We are all called to preach the good news in word and deed. List ways you do this. Pray to the Holy Spirit to help you.

God's Word In My Life

Holy Spirit . . .

"And they were all filled with the holy Spirit."

Acts of the Apostles 2:4

27
The Early Christians
ACTS OF THE APOSTLES 2:42–47, 4:32–35

Bible Background
On Pentecost, many people were baptized as Christians. The community of believers began to grow. They met in their homes. They did not have churches yet. Jewish people who followed Jesus still prayed in the Temple.

Reading the Word of God
Read the **Acts of the Apostles 2:42–47** and **4:32–35.**

Find out how the early Christians live.

Understanding the Word of God
Luke tells that those who follow Jesus are called followers of the Way. Jesus is the "Way." Believers live the way Jesus taught them. They share or sell all the things they own—their land and their homes. The followers spread the good news of Jesus. They break bread together as Jesus did at the Last Supper.

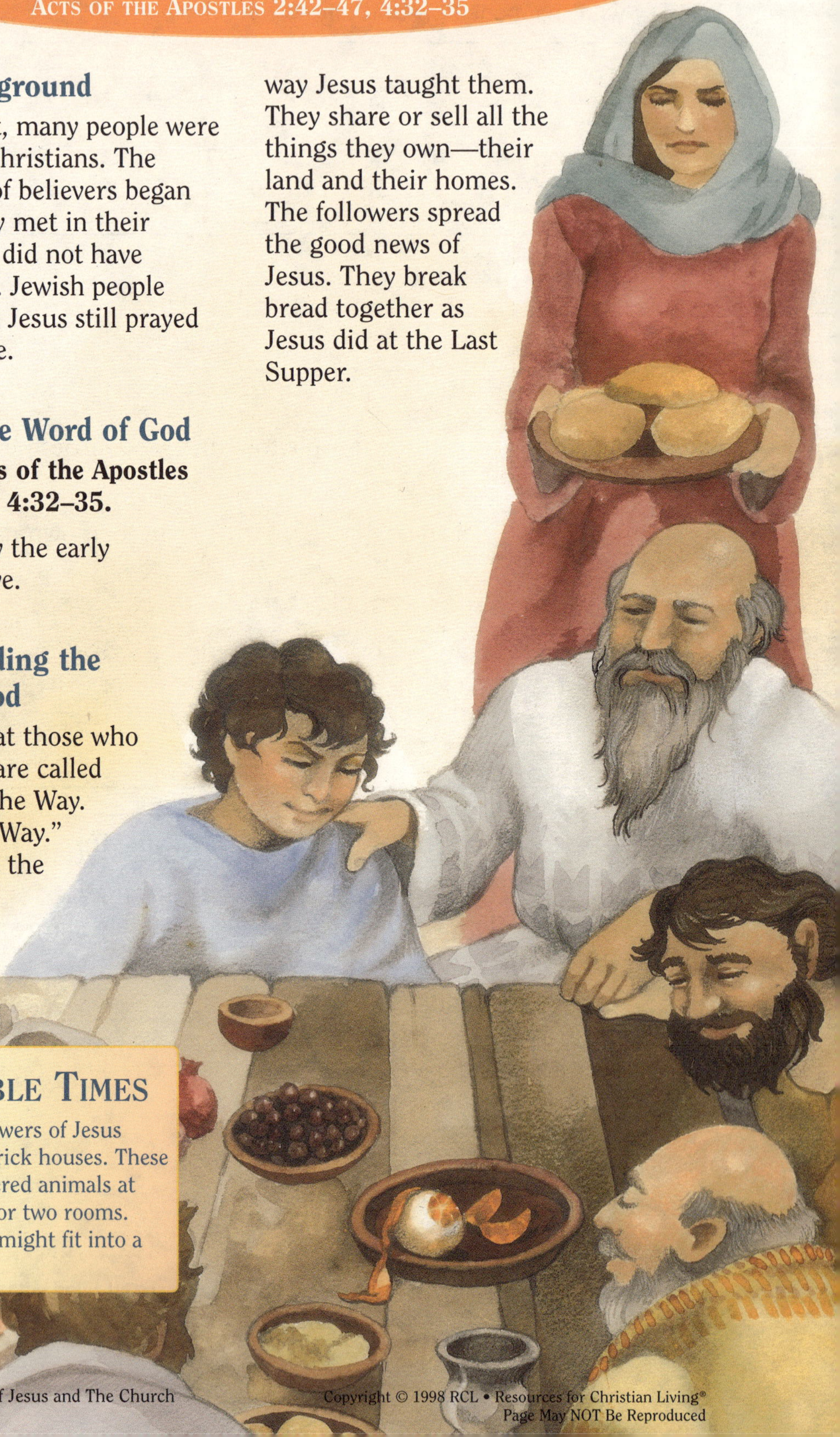

IN BIBLE TIMES
Many of the followers of Jesus lived in mud-brick houses. These houses, which sheltered animals at night, had only one or two rooms. All their belongings might fit into a chest.

Using the Bible to Write a Letter

The followers of Jesus lived together as a community of joyful believers. Reread the two summaries from the Acts of the Apostles. Write a letter as an early Christian. Explain how you live as a follower of Jesus.

Dear ____________________,

____

____

____

____

____

____

In the name of Jesus,

____

With Family or Friends

List ways to show how to live as a follower of Jesus today. Share the good news.

God's Word In My Life

As a Christian . . .

"Be glad in the Lord and rejoice, you just; exalt, all you upright of heart." Psalm 32:11

Scripture Glossary

Old Testament

Terms

Bible—the written Word of God

covenant—the faithful, loving commitment God made with God's people

Law or Torah—the first five books of the Old Testament

Old Testament—the first main part of the Bible. It tells us about the People of God who lived before Jesus was born.

Prophets—the books in the Old Testament written by the prophets

Sacred Scripture—the writings of the Bible, the written Word of God

Writings—books of the Old Testament

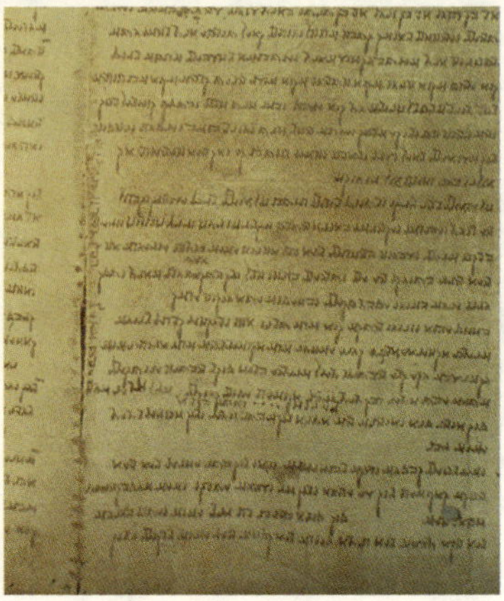

This "page" is from the Isaiah scroll. Stitching forms the pages into a scroll.

People

Abraham—the first great leader chosen by God to lead God's people

David—the name means *"beloved;"* a shepherd boy who was chosen by God as one one of the kings of God's people

Esau—the eldest son of Isaac and Rebekah, and the twin brother of Jacob

Hannah—Samuel's mother

Isaac—the only son of Abraham and Sarah

Israelites—the people of Abraham chosen by God

Jacob—the son of Isaac and Rebekah, and the twin brother of Esau

Jonathan—the son of Saul

judges—leaders of God's people before there were kings

Naomi—the mother-in-law of Ruth

Rebekah—the second wife of Isaac, and the mother of Esau and Jacob

Ruth—a kind woman who chose to leave the land of her birth to take care of her mother-in-law

Samuel—one of the judges of Israel. He was called by God to lead the people.

Sarah—the wife of Abraham, and the mother of Isaac

Saul—the first king of God's people

Places

Canaan—the place where God made his covenant with Abraham

Egypt—a country along the Nile River. The Israelites were held there as slaves.

Israel—the name means *"let God shine;"* the name given to Jacob; the land of God's chosen people

Moab—the land east of the Dead Sea to which the Israelites traveled to avoid famine

Abraham lived in Haran before going to Canaan. These unusual houses are located near Haran in southern Turkey.

MEDITERRANEAN SEA
CANAAN
ISRAEL
MOAB
DEAD SEA
GOSHEN
EGYPT
Nile River
Sinai Peninsula
MIDIAN
SEA OF REEDS (RED SEA)
Mt. Sinai

Scripture Glossary

New Testament

Terms

Acts of the Apostles—Luke's writings in the New Testament that tell about the work of the Holy Spirit and the beginning of the Church

Ascension—when Jesus returned to his Father in Heaven

Benedictus—Zechariah's canticle that "blesses" God

canticle—song of praise to God

Crucifixion of Jesus—the nailing of Jesus to the cross on which he died

Gospels—the first four books of the New Testament that tell us who Jesus is and what he did

Last Supper—the special meal Jesus and his disciples ate together before he died. At the Last Supper Jesus gave us the gift of his Body and Blood.

Magnificat—Mary's canticle, Mary's song

Messiah—word meaning *"anointed one;"* the Anointed One promised by God to save God's people. We believe that Jesus is the Messiah promised by God.

miracle—a sign of God's power and presence with us

New Testament—the second main part of the Bible. It tells us who Jesus is, what he did, and what he taught.

parables—stories used by Jesus to teach a lesson

Passover—a Jewish feast celebrating the passage from slavery in Egypt to freedom

Pentecost—the day that celebrates the Holy Spirit coming and living with us

Rabbi—word meaning *"teacher;"* the name given to the scribes by their students

resurrection of Jesus—the rising of Jesus from the dead

Revelation—the last book of the Bible. It is God's message to the seven Churches in Asia as revealed to John.

Sabbath—the seventh day of the week in the Israelite and Jewish calendar; a holy day; a day of rest

Places

Bethlehem—a town called the "city of David;" the birthplace of Jesus

Emmaus—a village near Jerusalem. On the day Jesus resurrected, he spoke with two of his disciples on the road to Emmaus.

Garden of Gethsemani—the place where Jesus went to pray after the Last Supper; the place where Jesus was arrested

Golgotha—the place where Jesus was crucified

Jerusalem—a city in Palestine that was called the "city of God." It was the center of religious life for the Jewish people.

Mount of Olives—the place where the Garden of Gethsemani was located

Nazareth—a village of Galilee; the home of Mary and Joseph; the place where Jesus grew up

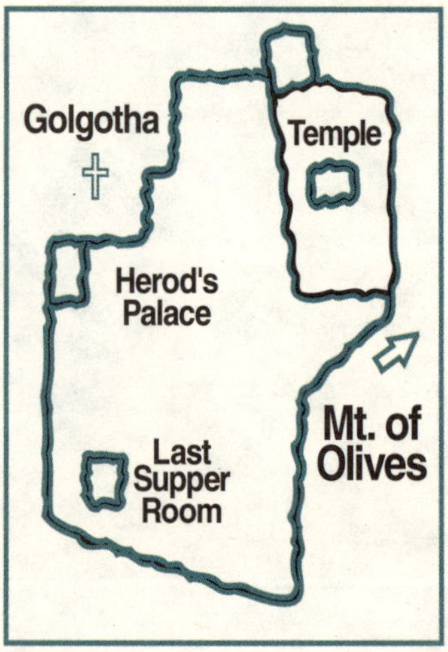

The Mount of Olives lies across the valley from Jerusalem. Experts think it marks the very spot where Jesus and his disciples went after the Last Supper.

Damascus
SYRIA
PHOENICIA
Tyre
GALILEE
Capernaum
Bethsaida
SEA OF GALILEE
Mt. Carmel
Cana
Nazareth
Mt. Tabor
MEDITERRANEAN SEA
Caesarea
SAMARIA
Jordan River
Samaria
Mt. Gerizim
Joppa
Jericho
Emmaus
Jerusalem
Bethany
Qumran
JUDEA
Bethlehem
DEAD SEA
Hebron
Gaza

This coin was minted by Rome about the time of Jesus' birth.

People

Anna—a prophet who recognized the infant Jesus as the Messiah

apostles—the twelve leaders Jesus chose and sent forth to continue the work he began. After Jesus' ascension, the Church named two more apostles.

disciples—those who answered Jesus' call and followed him

Elizabeth—the wife of Zechariah; mother of John the Baptist; relative of Mary the mother of Jesus

Gabriel—God's messenger to Mary who greeted her with the good news that she was to be the mother of Jesus

Holy Family—Jesus, Mary, and Joseph

Jairus—a religious leader of faith

Jesus—the name means *"God saves;"* the Son of God, who became man

Joseph—the foster father of Jesus, and the husband of Mary

Joseph of Arimathea—a disciple of Jesus who buried the body of Jesus after his death

Judas—the apostle who betrayed Jesus

Martha—the sister of Mary and Lazarus

Mary—the mother of Jesus, the Son of God made human

Mary—the sister of Martha and Lazarus

Matthias—chosen to replace Judas as one of the twelve apostles

Peter—the brother of Andrew called by Jesus to be one of the twelve apostles. Peter was named by Jesus to be the head of the Church.

Pharisees—a group of Jewish leaders who lived strictly by the law

Pilate—the Roman governor who turned Jesus over to be crucified

Simeon—a holy man in Jerusalem who recognized the infant Jesus as the Messiah

Zechariah—the father of John the Baptist

Scripture was written on scrolls. Scrolls were stored in jars.